HOW WE BUILD

ROADS

Philip Sauvain

M

Schools Library and Information Service

First published 1989

Published by Macmillan Children's Books
A division of MACMILLAN PUBLISHERS LTD
Houndmills, Basingstoke, Hampshire RG21 2XS
and London
Companies and representatives
throughout the world

British Library Cataloguing in Publication Data
Sauvain, Philip
    Roads. — (How we build)
    1. Roads. Construction
    I. Title II. Series
    625.7.

ISBN 0-333-45404-9

Designed by Behram Kapadia

Printed in Italy

Photographic acknowledgements

t = top b = bottom l = left r = right

cover: J. Allan Cash

6t The Hutchison Library; 6b Tony Stone; 9 C. M. Dixon; 11 Behram Kapadia; 12l Behram Kapadia; 12r ZEFA; 15 Gordon Carr; 15 [inset] Tony Stone; 16t ZEFA; 16b Gordon Carr; 18 Peter Carmichael/Aspect Picture Library; 22-23 J. Allan Cash; 24 J. Allan Cash; 26 J. Allan Cash; 27 Costain Ltd; 28 Derek Bayes/Aspect Picture Library; 29t Stenoak Fencing; 29b The Photo Source; 30t Behram Kapadia; 30b Simon Warner/ZEFA; 31t J. Allan Cash; 31b J. Allan Cash; 33t J. Allan Cash; 33b Behram Kapadia; 34 Costain Ltd; 35t Behram Kapadia; 36 G. Tompkinson/Aspect Picture Library; 37 J. Allan Cash; 38 The Hutchison Library; 40 Tony Stone; 41 The Hutchison Library; 42 General Motors; 44l Tony Stone; 44r Guinness Publishing; 45t M. Mann/Vision International; 45b G. Tompkinson/Aspect Picture Library

Note to the reader
In this book there are some words in the text which are printed in **bold** type. This shows that the word is listed in the glossary on page 46. The glossary gives a brief explanation of words which may be new to you.

Contents

Why we need roads

How far could someone drive across a country if there were no roads? It would depend on the nature of the land, of course, but it probably would not be very far. Fortunately, **engineers,** have found ways of laying roads across sandy deserts, through mountains and swamps, in steaming, tropical forests and in frozen arctic regions.

Different types of road

A road is a strip of ground suitable for a vehicle to travel on. It differs from the land on either side because its surface is hard enough to bear the weight of traffic.

One road may be very different from another. In remote places like the

▲ Roads through tropical rainforest, like this one being built in the Amazon rainforest, get muddy when wet, and dusty when dry. They are, however, main roads for many people.

▼ These busy roads in Los Angeles, in the United States, have a hard surface. It will stand up well to wear, despite the thousands of vehicles that drive over it every day.

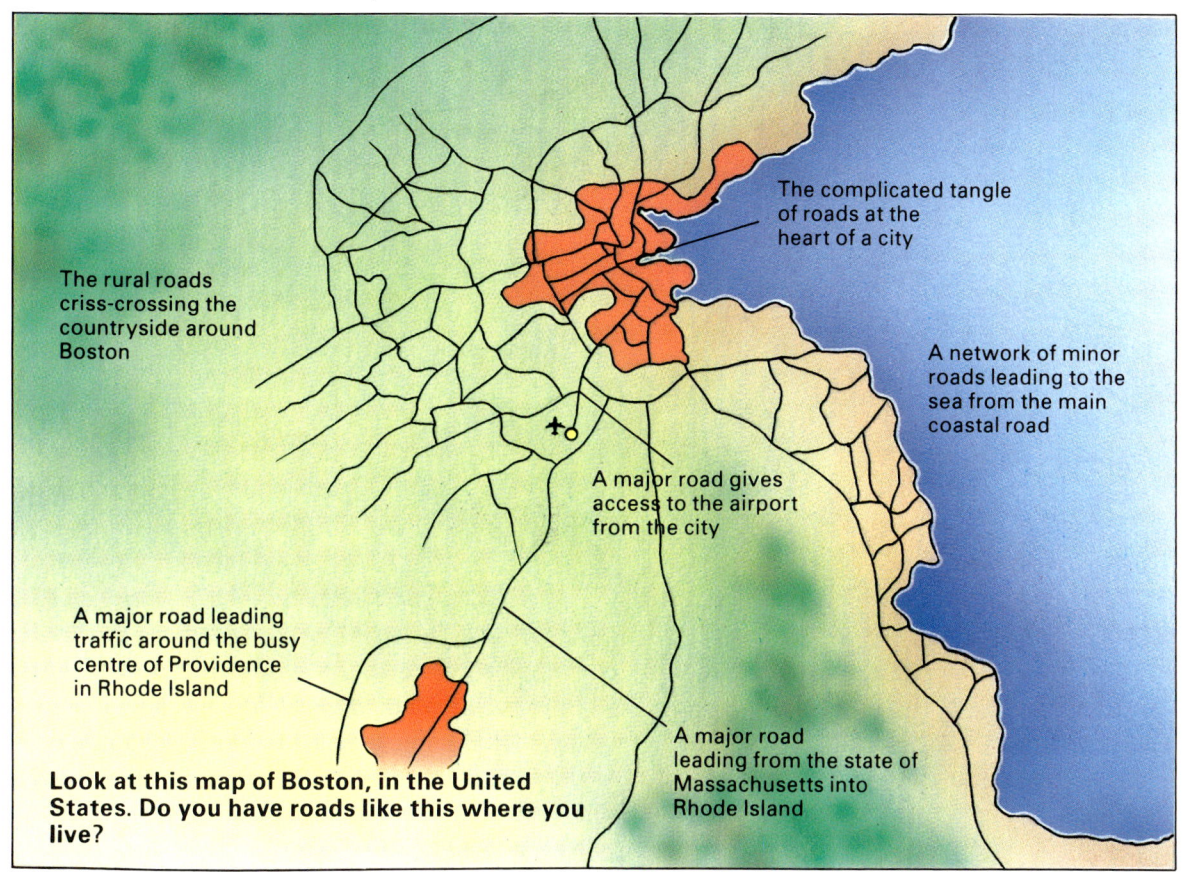

The complicated tangle of roads at the heart of a city

The rural roads criss-crossing the countryside around Boston

A network of minor roads leading to the sea from the main coastal road

A major road gives access to the airport from the city

A major road leading traffic around the busy centre of Providence in Rhode Island

A major road leading from the state of Massachusetts into Rhode Island

Look at this map of Boston, in the United States. Do you have roads like this where you live?

Himalayan mountains in southern Asia where cars are rare, roads are often just made of flattened earth or paved with loose stones. In rural areas, a narrow road may be wide enough to carry the few cars or tractors that use it. However, the roads into a big city carry many lanes of traffic.

The benefits of roads

Roads help people to keep in touch with each other by providing a direct route from one place to another. Roads are also used to supply many of our needs. Trucks bring farm produce to market and raw materials to factories. Many more products are available in areas with good roads than in isolated areas. In fact, most of the food we buy has been delivered to the shops by road. Road tankers supply fresh milk to our retailers and petrol to our local filling stations.

The drawbacks

Our way of life today depends heavily on roads. However, there are drawbacks as well due to the increased use of roads. Many people are injured or killed in accidents every year. Road traffic is often noisy and exhaust fumes from motor vehicles pollute the air we breathe.

As you read this book, you will learn more about roads. You will see how they are built and find out about the problems engineers have to solve in order to build them.

The first roads

People have used trails and tracks for their hunting and trading from the beginning of time, but the first paved roads we know of were built in Egypt and China, between 3000 and 2000 BC. The greatest road builders of the ancient world, however, were the Romans.

About 2000 years ago, after the Roman conquest of large parts of Europe and North Africa, thousands of slaves were put to work helping the Roman army to build 8000 kilometres of roads throughout the Roman empire. Roman roads stretched from Scotland across Europe to Jerusalem in the Middle East and beyond. Soldiers on foot and on horseback marched and rode along these roads. Supplies were carried on heavy wagons, while government officials travelled in horse-drawn vehicles. Mail, food, slaves, trade goods and treasure from all parts of the empire were sent back to Rome by road.

The Roman road system contributed to the success of the Roman Empire. The network of roads spread throughout the empire, which meant it could be more easily controlled.

Building the road

All Roman roads were built in the same way. First, the builders dug into the ground and fitted large stones together on top of the area they had cleared to give the road a strong foundation. They covered these stones with a layer of gravel and sand so that rainwater would drain through the road and prevent it becoming waterlogged. Then flat stones were laid evenly on top to provide the

surveying with a groma

clearing the forest

digging a ditch at the side of the road to drain away water

wearing surface

camber

laying paving stones

gravel and sand

▲ Roman roads were made wide enough so that the legions could march along them in columns which were six soldiers wide. The remains of this Roman road are on Wheeldale Moor in Yorkshire, England.

wearing surface, the part of the road that gets worn down by the pounding of wheels, hooves and feet. The road surface was given a **camber,** which meant that it sloped slightly from the middle to the sides, allowing rainwater to run off into the deep ditches that were dug at either side.

The Roman road engineers used an instrument called a groma to keep their roads as straight and as level as possible. Instead of building their roads around hills, they often built them straight up one side of a hill and down the other.

Today, the remains of Roman roads can be seen in many parts of the world. The Romans were expert road-builders and the fact that many of their methods are still in use is a reminder of their engineering skills.

◀ The work needed to build a Roman road was done by manual labour, not by machines.

Build a Roman road

You will need: a rectangular plastic tray, cardboard, modelling clay, sand, stones, water and pebbles of different sizes. Use the illustration as your guide.

1 Lay a road down the centre of the tray using the stones. Make the middle very slightly higher than the sides.

2 Fill in the spaces with pebbles and spread a layer of sand evenly on top.

3 Cut flat paving stones out of the clay and lay them on top, fitting them together as closely and smoothly as possible.

4 Cut two strips of cardboard the length of the tray and about five centimetres wide, and fold them lengthwise. Place them in either side of the tray to make the drains.

5 Pour some water on the road you have made. What do you notice?

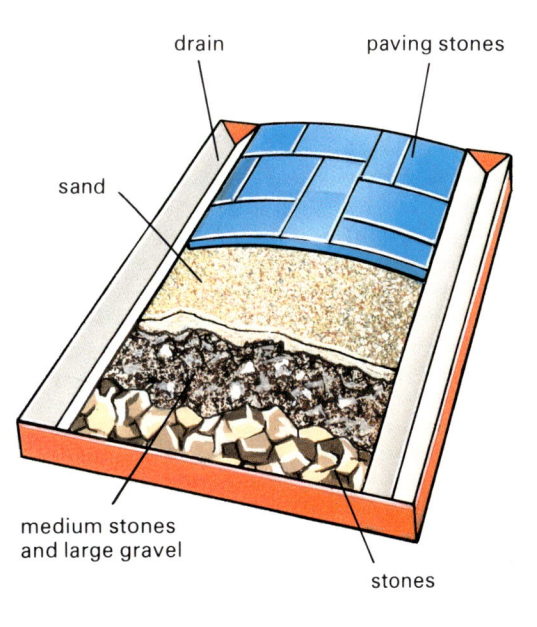

drain   paving stones

sand

medium stones and large gravel

stones

Coach roads

After the fall of the Roman empire, about 1400 years ago, most of the fine Roman roads fell into disrepair. Without the unifying power of Rome to maintain communications and to protect travellers, the movement of people and of armies became greatly restricted. People rarely travelled outside their villages. When people did travel, they made their journeys on foot or on horseback, using the old tracks and paths, and carrying their goods on the backs of pack animals.

Taxing the road users

Good roads were not built again until the end of the 1600s, when more and more people were travelling by stagecoach. In Britain, in 1663, an Act of Parliament was passed requiring travellers to pay for the use of roads at gates, called turnpikes, placed along the route. Countless coaches carrying mail and passengers passed through these gates each year. The payment, called a toll, made it possible to repair existing roads, and then, as wheeled traffic increased, to build new ones.

In the 1700s, road builders like John Metcalfe and Thomas Telford in Britain, and Pierre Tresaguet in France, built many excellent roads, while John McAdam developed new methods of making foundations and surfaces for roads, that are still in use.

The first turnpike road in the United States was built in Pennsylvania in 1793, as Americans began to move westwards. Sixty years later, stagecoaches carrying passengers, goods and mail were a familiar sight in the American West.

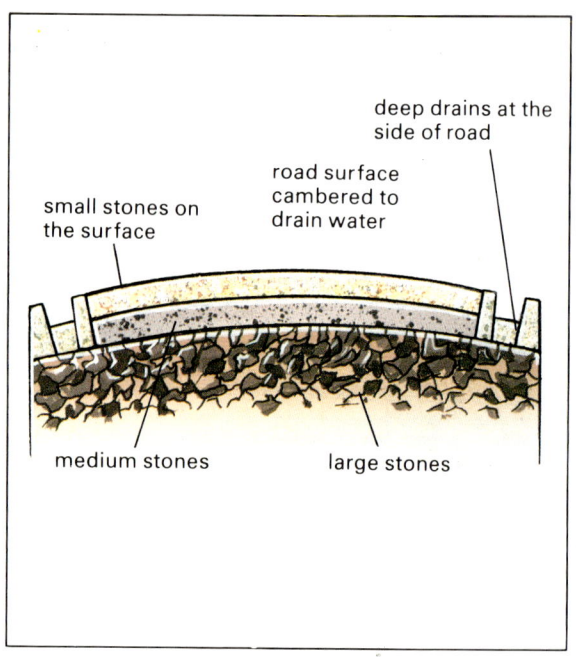

▲ Thomas Telford built his roads like Roman roads, with strong foundations made of large stones.

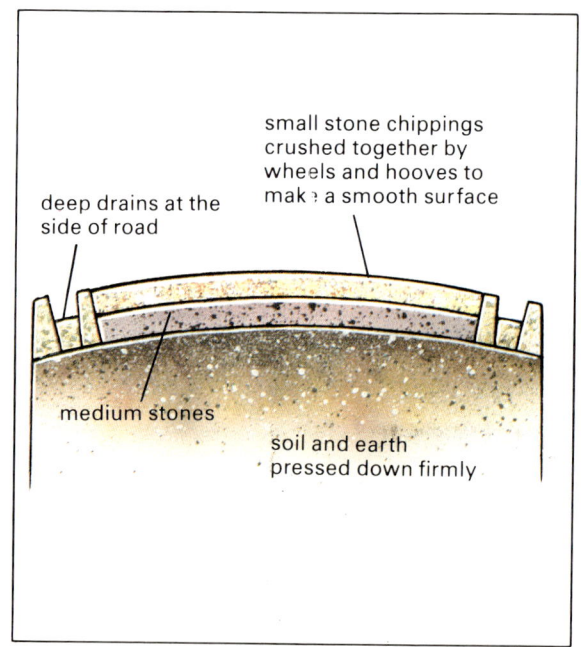

▲ John McAdam put stone chips on his roads. The wheels of carriages, wagons and coaches pressed them into the surface and made it hard.

turnpike or tollgate   turnpike keeper   tollhouse   guard   coach driver   mail coach   stagecoach   turnpike road

▲ A British turnpike road about 160 years ago. Fresh horses were harnessed to the stagecoach, and passengers stopped for food or rest at coaching inns along the way.

Motor roads

In the 1800s, the railways became the fastest, most dependable means of transport. Roads did not regain their importance until the coming of the motor car in the early 1900s. Then, it was found that the roads used for horses were not suitable for cars. The roads were too narrow and muddy, and motorists grumbled that horseshoe nails which had been left on the road surface, often burst their tyres. The wheels of the new 'horseless carriages' threw up clouds of dust which made it hard for the choking drivers to see the way ahead. Cars needed roads with a hard, smooth surface made from tar or concrete. It is from this point that the present world-wide explosion of highly-skilled and technical road engineering began.

▼ This poster is over 75 years old. At the beginning of this century, there were not many motor vehicles on the road.

Types of road

The main difference between one type of road and another lies in the frequency of its use within the road network. Is it an essential route between cities and much travelled on? Or is it a road between small towns, with a much lighter load of traffic?

Minor roads

The least used roads are the side streets in cities and the side roads that branch off from main roads in the country. They serve a comparatively small number of people. Often, these smaller roads are narrow and poorly surfaced. To use them, cars and trucks must go slowly, as these roads are not built for speed.

Major roads

Major roads vary in the volume of traffic they carry. There are the long distance major roads with their many lanes and crossing points, or intersections, down to the dirt roads that may be found in remote regions. They are often classified on road maps as primary roads, because they are the best roads available. Secondary roads take slower short-distance traffic.

The approach to primary roads is usually controlled by the use of specially-designed approach roads, traffic lights, or roundabouts, so that drivers coming in from secondary roads do not interrupt the flow of traffic. Such roads are often divided with a **central reservation** in the middle to make sure that vehicles going in opposite directions do not meet.

▼ A narrow street in the south of France.

▼ A German autobahn packed with vehicles in the middle of the rush hour.

Road systems

There are numerous names for the systems of major roads built in recent years to link our cities. In North America, they are called freeways, expressways, thruways, and even turnpikes. In Britain they are called motorways. In countries like the United States, Canada and France, drivers pay a toll to drive on some of these roads. Other major roads pass over or under them, joining them only by way of carefully planned **access roads.** A network known as the interstate highway system connects all the major cities in the United States. Most countries have, or plan to have, a similar network of fast, modern major roads.

▾ Getting onto local roads and city streets is easy, but you can only join a major road at certain places. Some road users, like cyclists, are not permitted to use them at all.

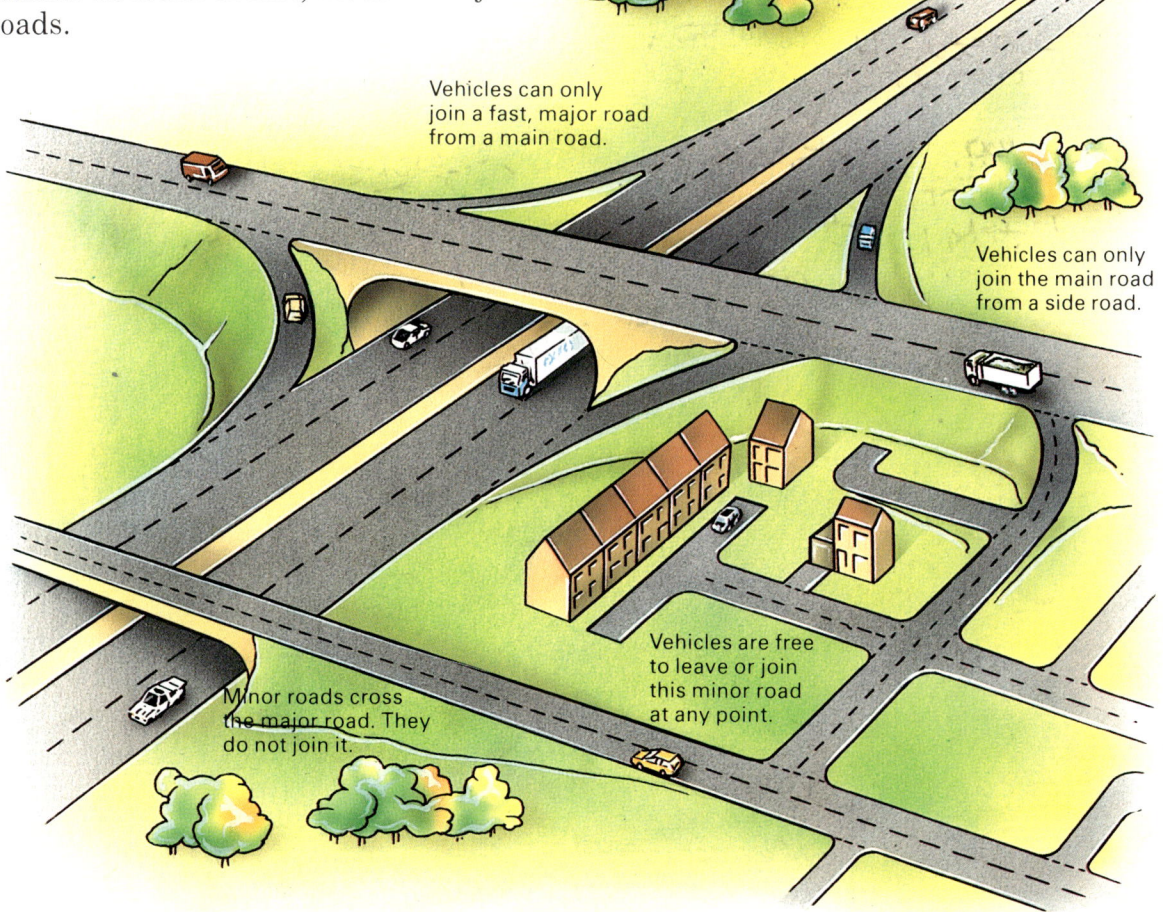

Vehicles can only join a fast, major road from a main road.

Vehicles can only join the main road from a side road.

Vehicles are free to leave or join this minor road at any point.

Minor roads cross the major road. They do not join it.

Planning a road

Have you ever been in a car that was stopped by an official asking questions? Where have you come from? Where are you going? A survey like this is called a traffic count. The questions are asked so that the people planning to build a new road can find out if it is really needed. From the answers they get, they can estimate how many vehicles will use the road every day if it is built.

Better roads

Many new roads are simply improved versions of the existing road, built to take more and heavier traffic along the same route. Other new roads are built to allow drivers to go around a city instead of through it. Besides being quicker for the motorist, a **bypass** helps to reduce congestion in the town's busiest streets. A road that encircles a city completely is called a ring road.

Where to build

First of all, the road planners have to acquire the land on which a road is to be built. Stretch a string between two points on a map and you will see the shortest, straightest route a road could follow. If it is possible to build it, that road would cost least in materials and labour. However, there may be artificial as well as natural obstacles. Buying and pulling down buildings is expensive and arouses strong feelings in the communities along the route. It may be cheaper to build a longer road to go around the buildings.

Road planners first draw the route of the road they are proposing to build on a

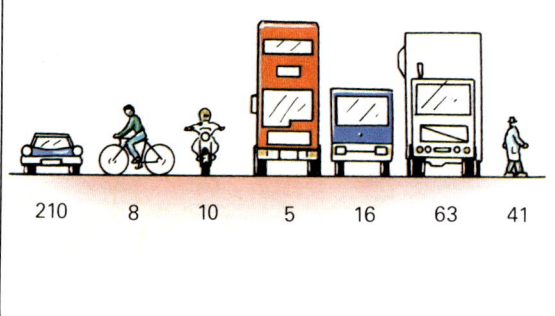
map and show it to the people concerned. The owners of the land or buildings may not wish to sell their property. However, if the road is being built with government approval, then, by law, the owners have to sell their property.

Public meetings are usually held to discuss plans for the new road. People may object to the road, fearing that it will spoil the countryside or put their children's lives in danger. Some may fear that the noise and fumes could harm wildlife in the area. Other people may welcome the business and work opportunities and the freedom from traffic jams that better roads bring.

▲ A designer at work, drawing plans and giving specifications for the road builders to follow.

▼ This motorway in England is under construction. Notice how the traffic is kept moving while the new road is being built.

Surveying the land

Before a new road is built, the road engineers survey the land it will cross using special instruments. You may have seen them using something that looks like a telescope on a tripod. This is a **theodolite.** The engineers use the theodolite to plan the direction in which the road will be built. They use an instrument called a level to measure how the land slopes and to make sure that the surface of the road will be on an even plane.

▶ A surveyor uses a theodolite to survey the course of a new road.

▼ An aerial photograph gives engineers a clear view of the landscape across which a new road has to travel.

Mapping from the air

Road engineers use pairs of nearly matching photographs, called **stereoscopic photographs,** of the proposed route of the road. These are taken from the air. Stereoscopic photos look like ordinary pictures until they are seen through a stereoscope, which makes the scene shown in the photos look three-dimensional. Engineers use these photographs to draw maps of the new road. Aerial photographs are especially useful to surveyors when they are planning roads through places like dense forest.

Maps drawn from the photographs show everything that could get in the way of the new road. They show the places where people live, as well as hills, rivers, lakes and trees. The engineers may have to build bridges or tunnels to overcome some of these obstacles.

Detailed plans

With this information, the engineers can now draw more detailed plans, in which the exact position of each part of the new road is clearly shown. These plans show how the road will bend, and where the pipes and small tunnels will be laid to take streams under the road. Meanwhile, other engineers are drawing up plans for the bridges and tunnels needed to carry the road across valleys and through hills.

Using a level

You will need: an empty, flat-sided, clear bottle with a screw top.

1 Fill the bottle with water right up to the top. Then screw the top on tight, making sure it does not leak.

2 Lay the bottle on the table, making sure it lies flat. If the table is level, you will see a bubble of air in the middle of the bottle. If the table slopes, the bubble will move to the higher end of the bottle.

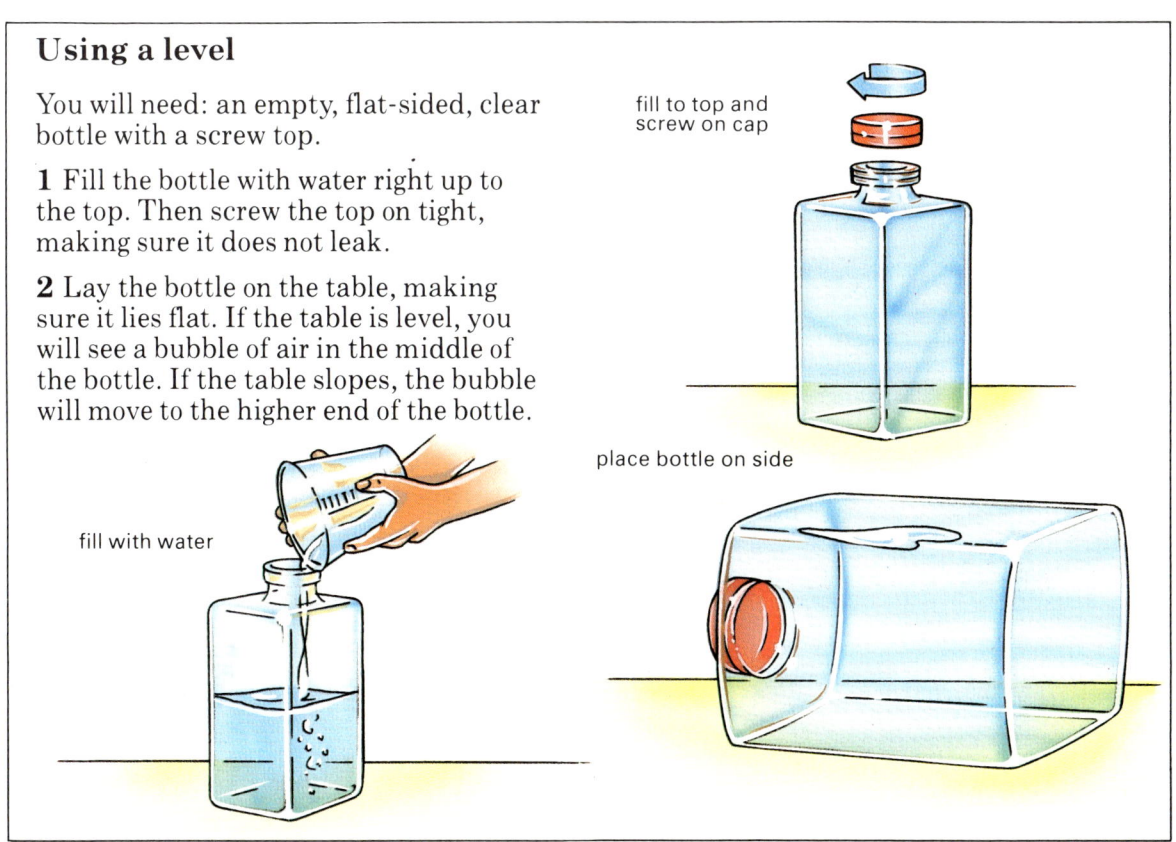

fill to top and screw on cap

place bottle on side

fill with water

Rocks and soil

Whatever the land is like along a planned route, it will be the new road's base, so it is important for the road engineers to know what kind of rocks and soils are there. The type of soil can affect the way in which the road is built. Sandy soils are the best for road-building, because they let water drain through them and stay the same in all weathers. Clay soils are the worst, because clay is waterproof and may prevent rainwater from draining into the ditches at the sides of the road.

▼ **Three bulldozers push huge piles of soil into heaps, so that it can be used to cover the banks at the sides of the completed road.**

Under the surface

Engineers sink **boreholes** along the course of the new road to take samples of the rocks and soils. They test these samples to find out what weight the soil can support without sinking. If the soil is soft, they may add gravel, small stones or sand to make the ground stronger. The boreholes also give the engineers information about the site which they could not have found out from stereoscopic photographs or from a printed map of the area.

The boreholes show the engineers whether there are any pits, tunnels or mine shafts below the course of the new road. The surface of the road could sink, or holes open up in it, if the remains of an old coal mine collapsed beneath it.

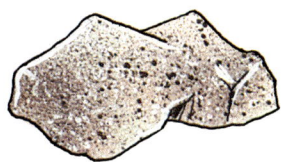

▲ Granite is a hard rock. Road builders often use granite chippings on road surfaces.

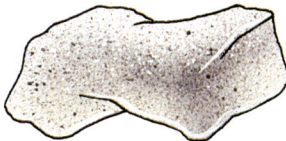

▲ Limestone contains lime, which is often used to make cement.

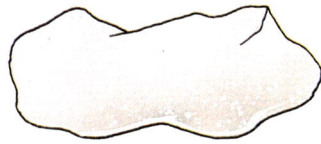

▲ Chalk is a soft white rock. It also contains the lime used to make cement.

▲ Road workers mix sand with cement to make concrete.

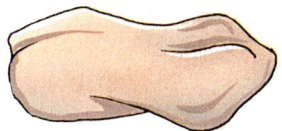

▲ Clay is hard when dry and soft when wet. Workers mix it with lime and sand to make cement.

▲ Gravel is small stones found in old riverbeds. It is used to build roads in many parts of the world.

Rocks build roads

Road builders use the rocks found in the area the road passes through for their building materials, if they possibly can. Using materials found on or near the site can greatly reduce building costs. Gravel is often found in heaps left behind by rivers thousands of years ago. Rocks and stones found in the area can be mixed with a black sticky material called asphalt to make the road surface. They may also be mixed with cement. Combined, they make concrete, which can be used either for the foundations or for the surface of a new road.

Testing soils

You will need: a trowel or an old spoon.

1 Go to several places where no one will mind if you dig up a handful of soil.

2 Take a sample in each place and test it. If you can shape the soil into a ball and it makes your hands dirty, it is a *clay* soil.

3 If the soil feels gritty, cannot be made into a ball, and leaves your hands clean, it is a *sandy* soil.

4 What types of soil are the roads in your area built on?

Note: most soils are part clay and part sand.

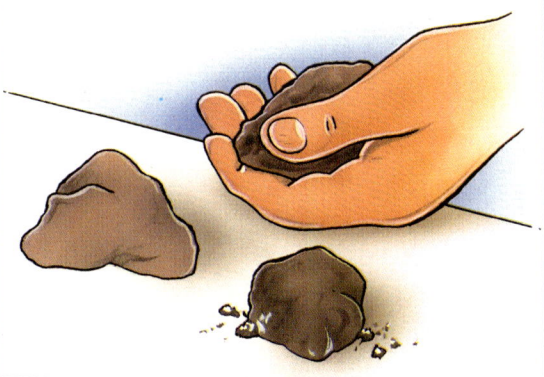

Draining the road

Falling rain sinks into the ground and fills up holes before it finds its way into streams and rivers. Engineers need to know how water drains off the land where a road is to be built. They check the annual rainfall and observe what happens after rain. Does the water drain away quickly? Do the rivers flood?

Planning against floods

So that flood water will be cut off before it can reach the roadworks or the completed road, the workers dig trenches called **cut off ditches** to catch the rainwater. Trenching machines are used to dig these ditches at either side of the route the road will take. Using pipe-laying machines, the road workers put large drain pipes into the ditches before covering them with gravel. When it rains, the water will sink through the gravel and into the pipes, which lead it safely away from the road. Small concrete tunnels, called **culverts,** carry the drainage pipes and pipes containing electric cables or telephone wires under the road.

Keeping the surface dry

If water is left standing in puddles on the surface of a road, it makes it difficult for drivers to stop their vehicles safely and can cause skidding. The road surface can be built with a camber so that water drains away down the sides of the road. The surface of a new road can be covered with a waterproof layer to prevent damage from rainwater.

▶ The drainage system of a new road. The culverts, pipes and tunnels keep water away from the road.

cutting in hill

road drains

embankment

stream carried
through a pipe
under the road

cambered surface
to let water drain
to the sides of
the road

drainage ditch

Moving earth

After all the tests and plans have been made, several teams of road builders each start to work on different parts of a new road at the same time. Their first job is to clear the site, using bulldozers to knock down small trees and to push the soil to one side. Bulldozers also clear away any buildings that are in the way of the new road. However, if buildings are to be preserved, moving equipment is brought in to transport them to a new site. Sometimes, rocks on the proposed route are so large that engineers have to use explosives to break them up. Sometimes, they have to dig or blast their way right through the side of a hill. These excavations are called **cuttings.**

Keeping the road level

When all the major obstacles have been cleared, the ground where the new road will be built is levelled roughly, or is **rough graded.** Huge grading machines or scrapers slice away the soil until the surface is level. Surplus soil cut away from one part of the road is used to fill up holes in other parts. This is known as the **cut and fill method** of road building. Sometimes, rocks, stones and earth are tipped into a deep hollow to make a slope, or **embankment.** The embankment will be planted with grass to make it look attractive.

Road engineers try not to move more earth than is necessary, because moving earth takes time and costs money. It is the biggest job road engineers do. On some roads, they move as much as a million tonnes of earth and rock for every five kilometres of road they build.

Grading machines and excavators move most of the earth and rocks. A strong blade under the grading machine slices through the earth as it moves forward, collecting it in a large bowl in the middle of the machine. When the bowl is full, the driver takes a load of over 50 tonnes of earth to the nearest embankment or dump. If the earth is needed a long way away, machines load it onto dump trucks for its journey, while the grading machine continues its work.

▼ **A road-building site is full of heavy machinery. Bulldozers and excavators are used to level the surface of the ground. A cut off ditch is being dug at the side of the road.**

A firm base

After the site of the road has been cleared and rough graded, the engineers give it a strong, firm base. The type of base they build depends on the amount and weight of traffic the road is expected to carry. A major road that is used every day by long-distance trucks and road tankers needs to have much stronger foundations than a narrow country lane.

Packing down the earth

Heavy machines called **compactors** pack and press the rough graded surface down hard to make it firm and solid. This layer of road is called the **subgrade.** The engineers make sure it will be strong enough to support the immense weight it will carry without sinking or crumbling.

Bulldozers with special compactor wheels smooth, scrape and roll the subgrade until it is strong enough to be used as a dirt road by the heavy earth-moving machines which are working on the site.

Hard layers

If no more work was done on the road, it would soon be damaged by rain or the weight of heavy traffic. To make the foundations durable, road engineers put several layers of hard material on top of the subgrade. A layer of crushed stones or gravel, about 30 centimetres thick, forms the sub base. Machines roll it hard and firm.

On some roads, the engineers also mix concrete with the subgrade to strengthen it. They may even lay a sheet of concrete on top of the sub base. This layer is called the lower base or road base.

Then, using a **paver,** the road workers cover the road base with a thin layer of hot tar and stones to make the upper road base. This material is called **macadam,** after the Scottish road-builder of the 1700s, John Loudon McAdam. The road is now ready for the final layer.

◀ A grading machine being used to prepare the subgrade for a new road.

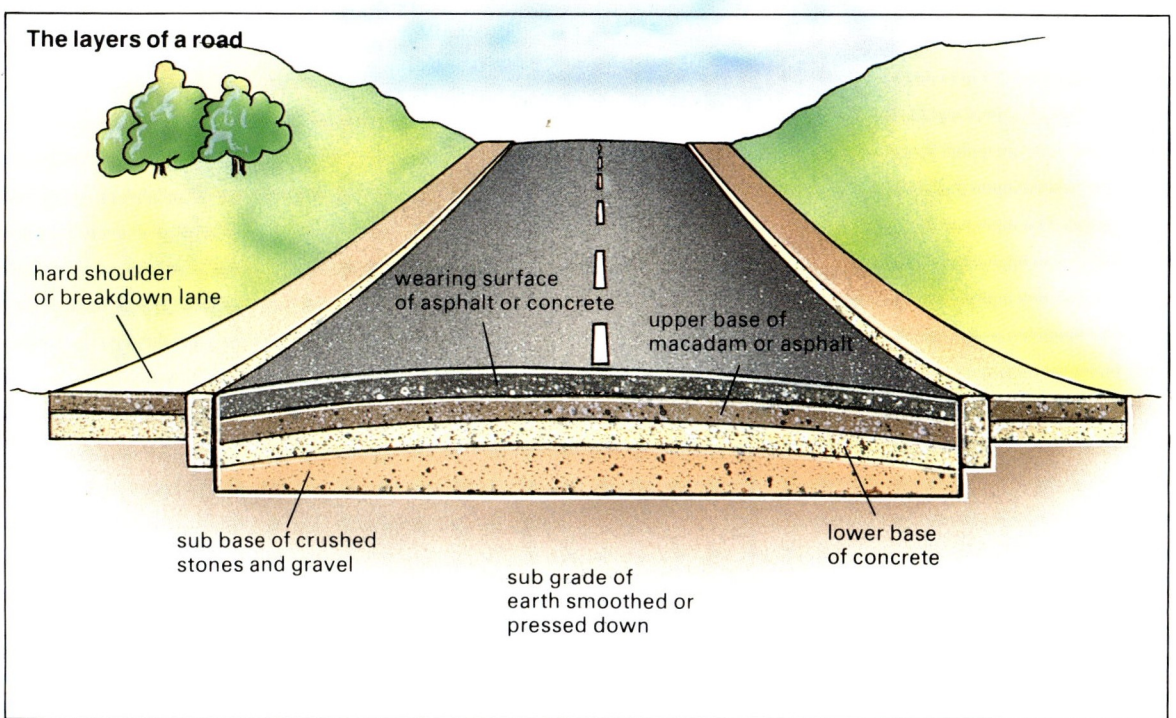

The layers of a road

hard shoulder or breakdown lane

wearing surface of asphalt or concrete

upper base of macadam or asphalt

sub base of crushed stones and gravel

sub grade of earth smoothed or pressed down

lower base of concrete

Build a road

You will need: a shoebox, five toothpicks, five small gummed labels and modelling clay in five different colours.

1 Make a model of the section of road shown in the picture. Use a different colour clay for each of the five layers, starting with a subgrade of brown clay.

2 Roll each lump of different coloured clay into pieces the size of stones, pebbles or gravel. Use them to make up the various layers of your model road.

3 Fold each label in half around a toothpick to make a flag.

4 Write the names of the layers on the flags and stick them in your model road.

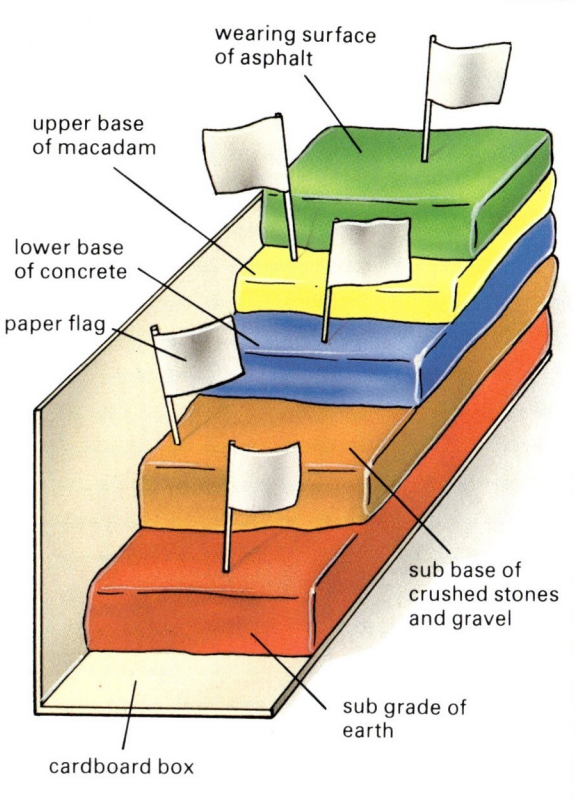

wearing surface of asphalt

upper base of macadam

lower base of concrete

paper flag

sub base of crushed stones and gravel

sub grade of earth

cardboard box

Laying the surface

The top layer, or wearing surface, of the road is usually made of either asphalt or concrete. Asphalt, or bitumen, as it is sometimes called, is a sticky black substance. When hot, it looks like thick black treacle, but when it cools, it sets

hard to make a firm, smooth surface. Drivers like **black top surfaces,** because they have a slight springiness which makes them smooth to drive on. They are flexible surfaces. Road surfaces made of concrete do not have that springiness. They are known as rigid surfaces.

▼ **Dump trucks and rollers lay a macadam road surface**.

A black top surface

To make a black top surface, the asphalt is heated on the construction site and mixed with small stones. Automatic pavers spread the hot mixture on the surface of the road. Several layers of asphalt may be needed, depending on the type of road that is being built and the amount of traffic that will use it. As the asphalt cools, rolling machines are driven back and forth over the surface until it is hard and smooth. Sometimes, stone chippings coated with bitumen are spread over the surface and rolled into it as well.

A rigid surface

To make a rigid surface, sheets of concrete about 20 centimetres thick are laid down by a series of machines called a **concrete train.** First, a concrete mixer combines cement with stones to make the concrete. Then, the mixture is poured into the spreader funnel on the front of the paver, which lays down an even layer of concrete between steel edges placed along both sides of the road. The edges prevent the concrete from spreading. A **skimmer** trims off any spare concrete. When the concrete has settled into place, steel mesh is rolled on top to form **reinforced concrete.** Then a second layer of concrete is laid on top of the first. The concrete sets hard to form the road surface.

When the concrete dries, it cracks. To make sure that the concrete cracks evenly, the concrete train puts a **crack inducer** into the concrete every six metres. Without the crack inducer, the concrete would crack unevenly. Then the finished road would not be able to carry the weight of traffic it was designed for.

Ready for traffic

Road-laying machines can lay up to three kilometres of road in a day. However, much additional building remains to be done before the road is ready for traffic.

No motorway or expressway can be declared open until the services which drivers need are also in place. They will want petrol stations, rest areas, and places to stop for meals. In these areas, plenty of parking space will be needed. Travellers may need picnic areas by the roadside, with tables and seating. Companies able to provide these services bid for the right to build or install them. In many countries, emergency telephones are installed along the sides of major roads, so that drivers can get help quickly if their cars break down or if they have an accident.

Road safety

Service areas may be provided for travellers, but it is up to the road engineers to ensure their safety. The engineers put up metal **crash barriers** in the middle of the road so that if a vehicle skids or goes out of control, it will not crash into oncoming traffic. They also put up warning lights and signals to prepare drivers for dangers ahead. They build a breakdown lane called a hard shoulder at the side of the road so that, in an emergency, drivers can pull over and stop.

The main lanes of the road must be clearly marked. Arrows and other symbols or words may be painted on the road to tell drivers when and where to slow down or change lanes. Good road signs are posted to notify drivers well in advance of an approaching **junction.**

▾ A petrol station on the Autoroute du Sud, in France. Behind the petrol station are shops and a restaurant.

Part of the countryside

When the road itself is finished, the road engineers usually do all they can to make the new road blend in with its surroundings. This is known as landscaping the road. The builders spread soil over the ground torn up by the construction work and plant grass, bushes, trees and wild flowers. Before long, the grassy banks at the side of the road and even the central reservation may again become a place where birds and plants and insect life flourish.

▼ Guard rails and warning signs help to make driving less dangerous on this scenic mountain road in Switzerland.

▲ Workers put up guard rails on either side of the central reservation as a dual carriageway nears completion.

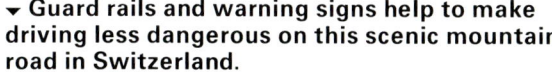

Road repair

Each time a car or truck uses a road, it wears down the surface by a tiny fraction. After several years of wear, a road may need maintenance, to keep it in good condition.

Road workers patch up small holes with asphalt. They dig out cracks in a concrete surface and fill the holes with fresh concrete or asphalt. Sometimes, the wearing surface of the road needs to be replaced completely.

The effects of weather

The weather does a lot of damage to roads. If water gets into a hole or crack in the surface, it may freeze. Ice takes up more space than water, so the effect is to loosen the road surface and make the hole bigger. The freezing and thawing can also cause the road to buckle, forming large bumps in the road. In some countries, salt and sand are scattered on roads to melt snow and ice. Although this makes driving safer, the salt and sand can damage the road surface.

Dirt roads quickly turn into mud in heavy rain. In hot, dry weather, the surface of a road often bakes hard and cracks open. Very hot weather can also melt and permanently damage the surface of an asphalt road. If the weather is hot during the day and bitterly cold at night, the change of temperature also weakens the road surface.

▲ Road repairs hold up traffic. Here vehicles using one side of a dual carriageway have to share it with vehicles from the other side. Road workers must place traffic cones so that it is immediately clear to drivers where they are to go.

▶ In many northern countries, snow may make roads impassable, but snowploughs like this one are used to clear the roads.

▲ Sandstorms and the drying heat make it difficult to maintain roads in desert country like this area in Algeria. Roadworkers have to clear the sand off the roads.

▶ Police, fire engines, ambulances and breakdown trucks are soon on the scene of an accident. Here a fuel tanker has overturned and the road has been closed to traffic because of the risk of fire.

Road care and safety

Good road maintenance makes roads safer. Roads need continual repair and rebuilding to make them safe in all weather conditions. Better lighting of roads at night reduces the risk of accidents. Traffic lights, bicycle lanes and pedestrian crossings in appropriate places also help. Even at their best, however, roads are dangerous places, and it is important for motorists, cyclists and pedestrians to follow the rules of road safety.

Safety

Today, roads are often built to speed up the flow of traffic. One way to do this is to build a bypass around a busy town. Another is to straighten out the bends. The fastest and safest roads are **dual carriageways** which carry traffic going in opposite directions well separated from each other. Motorists are forbidden to change direction by crossing into the oncoming lanes.

Slopes

Engineers can also speed up traffic by building roads that have a low gradient. A gradient is the distance a road goes forward in relation to the distance it climbs upwards. When engineers are building a road, they often cut the tops off hills and fill in valleys to cut down the gradient of the road.

Bends

Any sharp bend in the road slows down traffic. A motorist cannot drive at high speed around a sharp hairpin bend without risking a serious accident.

Road engineers design roads so that all vehicles lean over slightly as they go around curves. This is called **banking.** It allows drivers to steer around bends more safely.

Speed and safety

Every country has speed restrictions on roads and police to enforce them. The police are responsible for seeing that laws concerning speed, drunken driving, and the wearing of seat belts are observed. In some countries, slower vehicles like heavy trucks are forbidden by law to travel in the fast lane of a major road. Roads are safer if motorists drive carefully at reasonable speeds.

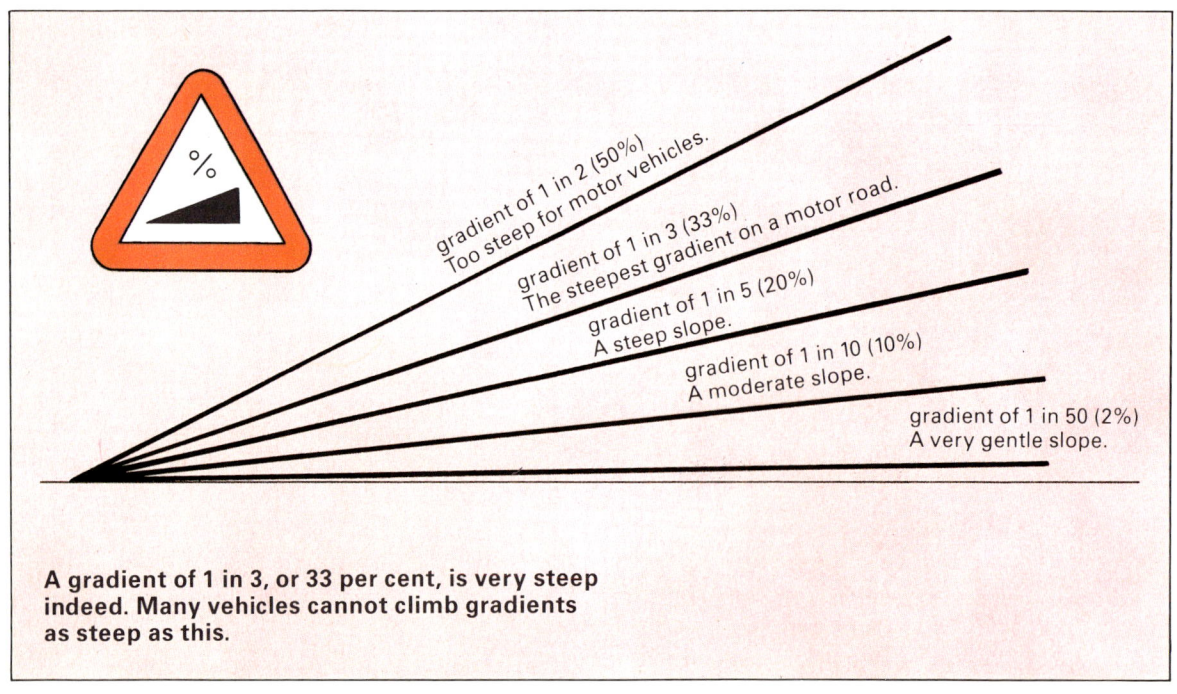

gradient of 1 in 2 (50%)
Too steep for motor vehicles.

gradient of 1 in 3 (33%)
The steepest gradient on a motor road.

gradient of 1 in 5 (20%)
A steep slope.

gradient of 1 in 10 (10%)
A moderate slope.

gradient of 1 in 50 (2%)
A very gentle slope.

A gradient of 1 in 3, or 33 per cent, is very steep indeed. Many vehicles cannot climb gradients as steep as this.

▲ You can guess why sharp curves like these are called hairpin bends.

▼ On this busy road in Britain, a slow lane has been built for heavily-laden trucks. Cars are not held up by the slower trucks.

Road crossings

There are a number of crossroads, or junctions on most main roads where drivers can turn right or left from one road into the other. However, crossing in front of another vehicle can be dangerous. It must be clear to drivers who has the right of way. Engineers may control the flow of traffic at an intersection by installing a roundabout or a set of traffic lights.

Major road intersections

Road engineers designing the intersections for major road systems make sure that the two roads never meet. Instead, one of the roads goes under the other. Drivers can usually only join one of the roads, or leave it to get on to the other road, by using special access roads. Drivers on an access road can only go in one direction, so four such roads are needed to allow traffic to get on and off both sides of the road. This system of interchanges makes these road systems safer than other roads.

Crossing points

When two major roads intersect, the crossing point can become quite complicated. The road engineers have to build at least eight access roads, and all of these roads must link up smoothly with one another. Drivers expect to change roads safely while still travelling at high speed. The simplest interchanges are diamond-shaped. Some look like a four-leaf clover. One place where major roads intersect, near Birmingham in England, is known as 'Spaghetti Junction'. This is because so many access roads weave in and out of each other at different levels that it looks from the air like a plate of spaghetti.

▼ **Two motorways pass over and under one another in this road system near Bristol in England.**

▲ A van drives along an access road to join this major road.

Major road intersections

A trumpet-shaped intersection

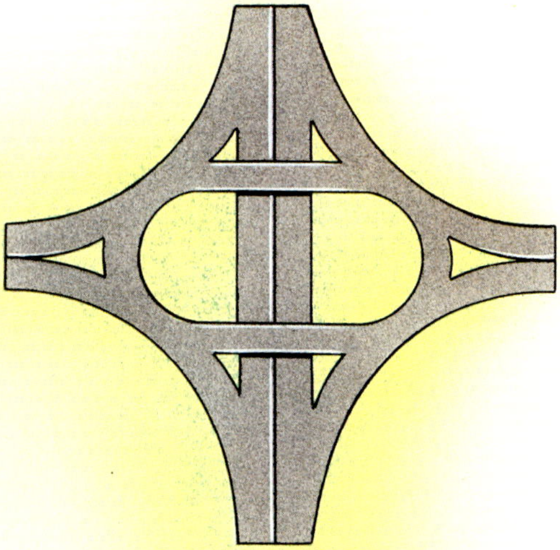

A roundabout intersection above a major road

A cloverleaf-shaped intersection

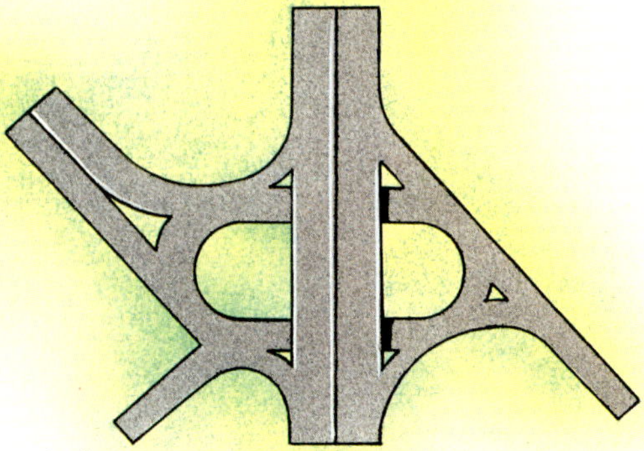

An intersection connecting a network of roads to a major road

Mountain roads

Road engineers have special problems to deal with when they build roads across high mountain ranges like the Alps in Europe or the Rockies in North America. A tunnel through a mountain provides the shortest route, but tunnels are costly to build. It is usually worth building a road tunnel only if it is on a busy route and will save motorists a lot of time.

▼ The Pan-American Highway, which starts in Alaska and ends in Chile, is shown here threading its way through the Andes Mountains in Peru.

Gaps in the hills

Most mountain roads wind their way through the valleys and gaps between mountains. The roads which climb up to the mountain passes often consist of a series of steep hairpin bends, which can present dangers to motorists.

All the difficulties of clearing and draining the site and cutting and filling and grading are multiplied when the road is constructed in high mountains. The huge machines which do so much of the work of laying the road have to be brought great distances. Sometimes they cannot be used at all because they are too heavy to be brought to the construction site.

Crossing a valley

Most of the Alpine towns and villages of Austria and Switzerland are situated in river valleys. Between them runs a good network of roads. Sometimes the road is carried high above the valley floor on top of arched bridges. The bridges keep the road on one level and

▲ A bridge takes the Brenner Pass Road straight across a deep valley amidst the spectacular scenery of the Austrian Tyrol.

greatly shorten its length. A bridge saves the work of blasting many kilometres of twisting and dangerous roads out of the rock of the mountainside.

Wet lands and forests

While mountains, rivers and valleys pose problems for road-builders, wet places, like marshes and swamps, present them with a different sort of challenge. The ground is probably very soft and flat. It may be flooded in wet weather and consist of deep layers of dust when it is dry.

In such conditions, it is vital that the road builders make sure that the new road has firm foundations and a good drainage system.

Cold, northern lands, like Alaska and Siberia, often have floods in the summer, because the soil is frozen just below the surface. This permanently frozen ground stops the rain from soaking away and makes it difficult to build roads. Where trees are plentiful, engineers sometimes lay branches across the ground and build the road on top of them. The Alaskan Highway, like many roads which have been built over frozen ground, is built on a deep sub base of gravel.

The Amazon highway

The hot and humid rainforest in many tropical countries also makes road building difficult. The ancient hardwood trees reach enormous heights. Road builders have to make their way through a dense undergrowth of fast-growing plants and creepers. The ground is often wet and marshy, and the rainfall is very heavy. There may be mosquitoes, snakes and crocodiles to contend with.

It was under conditions like these that road engineers built the roads which cross the Amazonian rainforest in South America. The building of the Trans-Amazonian Highway, though undertaken in the name of progress, destroyed large areas of rainforest and was a disaster for many of the people living in the forest, as it destroyed their whole way of life.

▸ **Road-builders often have to overcome a variety of natural obstacles.**

▾ **The Trans-Amazonian Highway runs through tropical forest in Brazil. The road builders have to clear the ground before they can lay the foundations of the road. Many of the trees are over 50 metres high.**

Planning a road

You will need: a sheet of tracing paper and a pen.

1 Study the picture below. A new road is to be built across this area. Imagine that you are the road engineer in charge of building it.

2 Which is the best route for the road? Where do you think it should go?

3 Trace all the main features of the picture. Then draw in your road. Show where you would build tunnels, bridges, embankments and cuttings.

The course of a road

Roads through cities

Many cities have been standing since long before motor cars were invented. As these cities grew larger and the number of cars coming into them increased, their roads needed to be constantly renewed, widened and modernised. Otherwise, they would not have been able to keep up with the demands made on them. Careful planning was, and still is, needed to avoid traffic jams at the times of day when workers from the suburbs are arriving in and leaving the city centre. Building and maintenance work must not interfere more than is necessary with the flow of traffic.

Urban roads

Special roads have been built in many cities in order to speed traffic through or past the heart of the city. Building urban roads is not easy because many city streets were first built for use by horse-drawn vehicles. They intersect at right angles, which means that traffic has to slow down at every corner. Road engineers cannot usually pull down buildings to make roads wider. To build a fast urban road, it is often necessary to lift the new road above the existing streets on columns or embankments to make an **elevated roadway**. Sometimes, they have to build tunnels to take the road under the busiest part of the city.

▼ **An elevated roadway makes it possible for motorists to drive quickly through a city.**

City problems

Speeding up the traffic into and out of cities makes it easier for people to get to or leave work. However, it may encourage people to bring their cars into the city instead of using trains or buses. Often, there is only one person in the car, yet each driver needs somewhere to park. Cars add to the noise, smell and traffic congestion in the city. Some cities have air that is seriously polluted by the smoke and fumes from factories, furnaces, cars and trucks.

The problems created by too much road traffic can often be eased, if not solved, by quite simple methods. Many of the problems can be avoided altogether, if care is taken on the design of new roads or traffic systems. Most cities have some one-way systems which can help the flow of traffic. Some countries have introduced special bicycle lanes and bus lanes. Some cities provide free car parks on their outskirts with a bus service into the city centre.

The future

What will the world's roads look like one hundred years from now? Will people still be driving cars on them? Some experts believe that all the world's oil will have been used up by then. The cars of the future may have to be powered by something other than petrol. They could run on electric batteries like fork-lift trucks or delivery vans do now. Electric cars have already been invented. Their main drawback is that they need heavy batteries which can only provide power for short distances before they need recharging with electricity.

▼ Cars like this solar-powered one would certainly change the look of our roads.

If the cars of the future are electric, we may not need roads. They could run on rails. Or, if the cars are built on the principle of a hovercraft, they could skim just above the ground. They might even be able to make their own power, running on electricity generated by the Sun. Cars powered by the Sun's energy have already been tested in Australia, Greece and Spain.

Changes to come

Even with these new developments, few people imagine that road transport as we know it is coming to an end. For the next 50 years, at least, we can expect to see our roads expanded and improved. Engineers may be able to develop new types of road surface. These could help tyres to grip the road better in bad

weather. Roads with more traffic lanes and separate roads for trucks will be built. Many more cities will have roads which will travel through tunnels underneath the city centre or elevated roads above the streets. Even in the country, roads may have to be built underground or above existing roads as the land on which to build them becomes scarcer.

Some day in the future, computers may do all the thinking and steering for the driver of a car, eliminating human error. This would mean faster and even safer travel. Future cars may even run on tracks like a train or tramcar.

Did you know?

* The largest ring road in the world is the M25 motorway in Britain. It is 196 kilometres long and encircles London completely.

* The steepest street in the world is Baldwin Street in Dunedin, New Zealand. Its maximum gradient is 1 in 2.66, or 38 per cent! Most people would find it hard to walk up a gradient as steep as this.

* The Pan-American Highway is the world's longest motor road. It stretches from northwest Alaska, in the United States, to Chile, in South America.

* If all the motor roads in the United States were laid end-to-end, they would stretch 175 times around the Earth.

* It would take a motorist driving non-stop along all the roads in the United States at a steady speed of 80 kilometres an hour, about ten years to cover the total distance.

* The longest road tunnel in the world is the St Gotthard Road Tunnel in Switzerland. It is 16.4 kilometres long and was opened in 1980.

* The world's biggest traffic jam on a main road occurred in 1980 in France. A line of vehicles 176 kilometres long came to a stop on the road between Lyons and Paris.

▼ The M25 motorway encircling London was opened in 1986, but in a few years it was already overcrowded during rush hours.

▼ Walking up Baldwin Street in Dunedin, New Zealand, can be an exhausting experience.

▲ The Dan Ryan Expressway in Chicago is one of the world's busiest roads. Over 250 000 vehicles travel along it every day of the week.

▶ Driving the full length of the Pan American Highway is an adventure which starts in Alaska and ends in Chile.

✳ The highest motor road in the world is in Tibet, China. At one point, the road reaches a height of 5632 metres.

✳ The first network of fast major roads was begun in Germany. The Avus Autobahn was opened in Berlin in 1921. By 1939, the German autobahns covered a total distance of over 4000 kilometres.

Glossary

access road: a one-way road leading onto or off a major road.

banking: making a road slope upwards from its inside edge on a curve.

black top surface: the name given to a road surface made of a black, sticky material which sets hard to make the road surface smooth and slightly springy.

borehole: a hole drilled in the ground in order to obtain samples of rocks and soil from beneath the surface.

bypass: a fast road that passes around a city.

camber: the slight slope from the centre of a road to the sides.

central reservation: a strip of land separating the two sides of a dual carriageway.

compactor: a rolling machine that is used to pack down the earth in the foundation layer of a road.

concrete train: a series of machines linked together to lay a concrete road surface.

crack inducer: the part of a concrete train that cracks the concrete at controlled intervals to prevent irregular cracking.

crash barriers: strong steel barriers erected to prevent vehicles leaving the lane or road they are on and causing accidents.

culvert: a drain or channel carrying water under a road.

cut and fill method: a method of using earth cut away from one part of the course of a new road to fill in a hollow in another part of the same road.

cut off ditch: a ditch or trench dug to catch rainwater and prevent it from flooding a road.

cutting: part of the course of a road or railway track cut through solid rock in high ground.

dual carriageway: a multi-lane road with the lanes going in one direction separated from those going in the other.

elevated roadway: a section of road which is carried on columns or pillars or on an embankment. Elevated roadways are often used to carry fast motor roads through built-up areas.

embankment: a mound of earth that has been used to build up a road or railway track to the same level as adjoining sections.

engineer: a person who makes use of scientific ideas in order to build structures.

junction: the place where two roads intersect.

macadam: a mixture of thick, black material such as coal tar and stone chippings, which is used for road surfaces.

paver: a machine that puts the upper layers of a road in place.

reinforced concrete: concrete that has had wire mesh or steel bars added to it.

rough graded: a road that has been cleared of obstacles and made level.

skimmer: a machine that trims away any concrete not needed for the road surface.

stereoscopic photograph: one of a pair of photographs taken from slightly different points which, when viewed together through a stereoscope, give a three-dimensional effect.

subgrade: the name given to the course of a road after it has been scraped and packed down to form a rough dirt road.

theodolite: the instrument used by a surveyor to measure directions and the height of an area of land.

wearing surface: the top surface of the road, which wears out first.